DON'T PICK ON ME

Dedicated to:

Michael K. M. Smith and Everton A.T. Smith

To the countless number of students, all around the world, who have been bullied or picked on in school or on the playground, on a team, in a family, or in their community.

To all adults who have had to deal with issues of bullying in the workplace, at home, or in the community.

All the amazing people: teachers, classroom helpers, social workers, and those that dedicate their time and service to improving the lives of young people in all capacities.

Acknowledgement:

Special thanks to my wonderful colleagues in the teaching profession, all the students and staff from Kane Middle School (Toronto, Ontario) & Smithfield Middle School (Etobicoke, Ontario), and the people without whom this book would not be possible including; Beverley, Andrea, Moses, Phyllis, and Arlene.

Don't pick on me because I am short

Don't pick on me because I am tall

Don't pick on me if I am a little skinny

Don't pick on me if I am a little chubby

Don't pick on me because I have a different
body type

Don't pick on me because of my gender

Don't pick on me

Just don't do it!

Don't pick on me because I am not as smart as you are
Don't pick on me or call me a nerd because I study hard
Don't pick on me because I struggle with some subjects
Don't pick on me because I am autistic
I may not speak like you or act like you, but I am me
Don't pick on me
Just don't do it!

Don't pick on me because I am not as
athletic as you are
Don't pick on me because I am slow, and
you are fast
Don't pick on me because I can't dribble a
ball like you can
Don't pick on me because I like a sport
that's not in your plan.
Don't pick on me
Just don't do it!

Treat others like you would
like to be treated.

Don't pick on me because of my customs
and traditions
Don't pick on me because of the God
I serve and my religion
Don't pick on me because of the way
I stutter
Don't pick on me because I don't have a lot
of friends.
Don't pick on me because I like to read
poems and giggle
Don't pick on me because... Yes! Just
because you should not
Don't pick on me
Just don't do it!

Don't pick on me because I'm in a wheelchair
Don't pick on me because I have a medical
condition
Don't pick on me because I wear glasses
Don't pick on me because I live with just my
mom or just my dad.
Don't pick on me
Just don't do it!

Don't pick on me because I don't wear
name brands
Don't pick on me because I can't afford
the things you can
Don't pick on me because I wear a turban
or a hijab
I'm not trying to offend you; that's how I
like it
I don't tell you how to dress or what to
wear
Don't pick on me
Just don't do it!

Don't pick on me because of the pigment
of my skin
Don't pick on me because of the colour
of my eyes
Don't pick on me because I am popular
Don't pick on me because I am quiet
and shy
It's tough just getting by from day to day
already
Don't pick on me
Just don't do it!

Black River
Public School

Don't pick on me because I live in a not-so-nice part of town
That's all we can afford now...
Don't pick on me because of the car my parents drive
It does its job just fine
Life is challenging, and we are doing the best we can
Don't pick on me
Just don't do it!

Don't pick on me because I come from a
land afar
Don't pick on me if my accent, to you,
sounds bizarre
Because yours sounds strange to me, too.
Don't pick on me because of the food
I eat
Don't pick on me because my music has
a different beat
Don't pick on me
Just don't do it!

Don't pick on me because you have a great
social media following
Don't pick on me because I don't follow you
on Facebook / Instagram / Snapchat
Don't pick on me because people are
spreading gossip or rumours about you
Don't pick on me because you think you are
cute or handsome
Don't pick on me because I don't really care
For I don't need special attention to know
who I am
Don't pick on me because I am the best me
I can be right now
And I am getting better day by day...
Don't pick on me
Just don't do it!

Don't pick on me because I am different in
my own way
Don't pick on me because I have money
and emotional issues
Don't pick on me because I am weak –
one day, I'll be strong
Don't pick on me because I am a geek;
with it, there's nothing wrong
Don't pick on me because I like to read
interesting books and sleep
Don't pick on me because you think I am a
freak
Say what you want, but leave me alone...
Don't pick on me
Just don't do it!

Our similarities and differences are what makes us unique

Don't pick on me, and I won't pick on you.

I have my issues, and you have yours

Leave me alone, and I will leave you alone

Be fair – treat me like you would like to be treated

Don't pick on me

Just don't do it!

Peace! I'm outta' here – be good – live good!

And, oh yeah, please DON'T PICK ON ME.

JUST DON'T DO IT!

FORMS OF BULLYING

There are many forms of bullying in our society.

Physical Bullying – this type of bullying starts as early as kindergarten. It can include biting, kicking, hitting, pinching, or pulling someone's hair. Someone may threaten to hurt you if you do not comply and do what they want.

Verbal Bullying - this type of bullying is sometimes as harsh as physical bullying without touching the victim. It can include insults, name calling, gossip, spreading rumors, and ongoing teasing.

Emotional Intimidation – this type of bully may deliberately exclude you from a group activity such as a game or not selecting you for a team or inviting you to an event.

Racist Bullying – this type of bullying includes making nasty racial remarks, spray painting graffiti, mocking the victim's cultural customs, and making offensive gestures. Often the bully, feeling empowered by their culture, tells you to go back to where you come from.

Sexual Bullying – this type of bullying is unwanted physical contact or abusive comments. It may also be verbal in nature making the victim feel uncomfortable.

Cyberbullying – this type of bullying is more common due to the influx of many social media platforms today. It includes all forms of social media including emails, Web sites, chat rooms, instant messaging and texting. It can be used as a form of threat, torment, harassment, humiliation, embarrassment or to target another person.

What to do when you are being picked on:

Tell your parent or guardian - they can teach young children about self-respect; to see the beauty they possess inside and help them to appreciate themselves for who they are.

Tell a teacher, vice-principal, or principal – there are policies and procedures for all school boards to follow to help to eliminate such treatments in a learning and cooperative environment.

Tell a trusted friend – a good friend will help you to stay strong in the face of adversity and may share advice or help you solve what may seem like an unending problem.

Tell a guidance counsellor / social worker : - they are another source of assistance – they can assist you or they may help to link you with external agencies to get the necessary support you require.

Reach out to support groups in your community – these advocate groups may offer advice and conflict resolution strategies.

For adults: It is important to share these ideas with your loved ones and seek help via your human resources department in your workplace. You may also refer to:

- Canadian Charter of Rights and Freedoms
- Human Rights Code of Ontario
- Discrimination under the Human Rights Code of Ontario workplace harassment

About The Author

Mike Smith was born on the lush tropical Caribbean island of Jamaica, in the parish of St. Andrew. He has been an elementary educator and teacher librarian for thirty years in Canada, primarily teaching Grades 6-8. He is passionate about reading, and books.

Mike resides in the Greater Toronto Area with his wife and two children. He is an avid musician and sports enthusiast. Don't Pick on Me is his first book, and it is related to the treatment of young people in the school system. Don't Pick on Me is aimed at starting a conversation about the bullying-related issues faced by pre-teens, teenagers, and adults alike. What solutions are available when someone is picking on you? What should you do? Where can you go to find help?

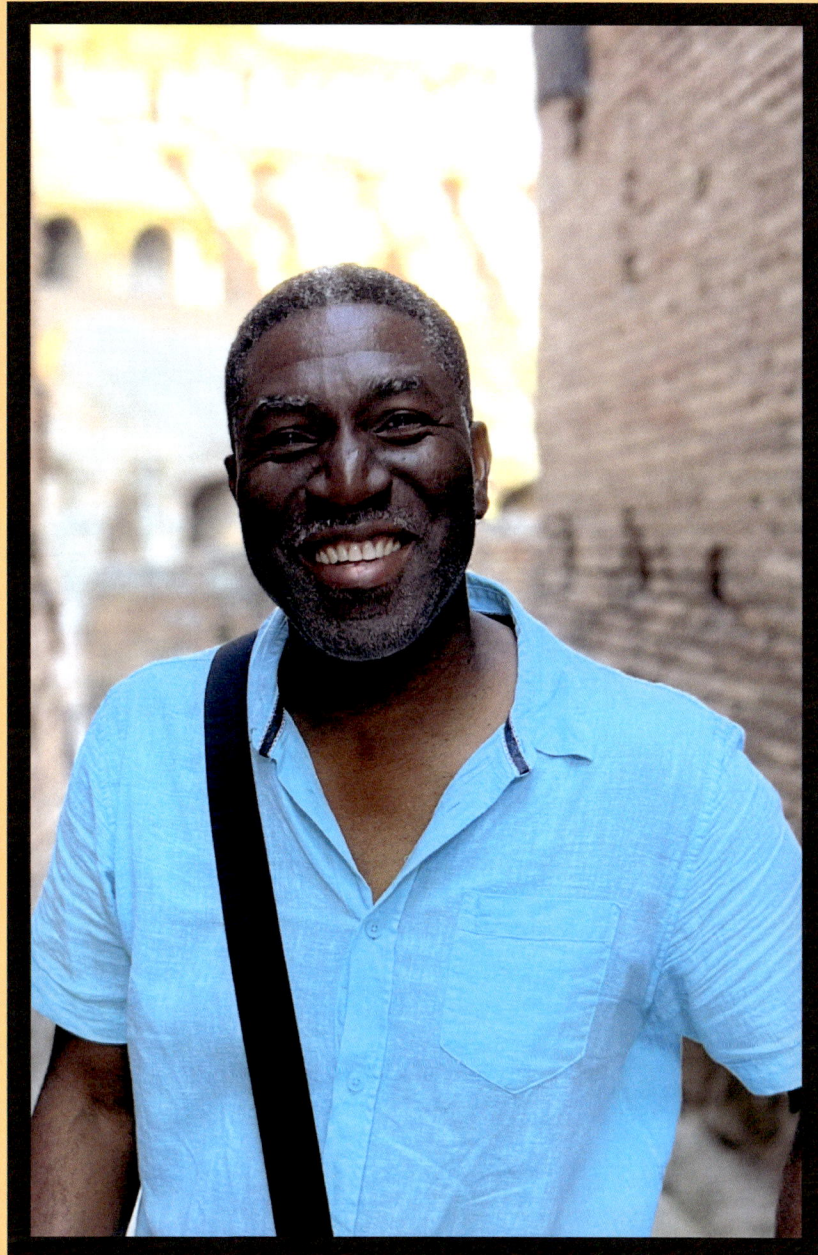

MIKE SMITH

www.ingramcontent.com/pod-product-compliance
Lightning Source LLC
Chambersburg PA
CBRC090851210326
41597CB00008B/163